Wanda Hears the Stars

Stars

A Blind Astronomer Listens to the Universe

Amy S. Hansen *with* Wanda Díaz Merced

Illustrated by Rocío Arreola Mendoza

iɔi Charlesbridge

Look up! ¡Mira arriba!

Can you see stars dance across the sky?

For Wanda Díaz Merced, the stars were hidden. Her family lived in Gurabo, a small town in Puerto Rico's rainforest.

Coquí frogs called.

KO-KEE

KO-KEE

KO-KEE!

Lizard cuckoos laughed.

Keh Keh Keh!

Crickets chanted.

chirrp chirrp!

The animals lived in the trees, and those same trees blocked the sky.

Early one morning when Wanda was nine,
she went with her family on a fishing trip.
Looking up she saw hundreds—*millions*—of stars.

They pulsed. They twinkled. They astounded her.

A meteor shower pushed into view. ¡Los colores!
Falling stars fizzled like sparklers.

Wanda felt something pop and shimmer inside her.

Her father spotted the meteors. "Esas estrellas son piedras que caen del cielo," he said. *Those stars are stones that fall from the sky.*

Wanda had questions. "¿De qué están hechas las estrellas?" *What are stars made of?* "¿Qué están haciendo en el cielo?" *What are they doing in the sky?*

Her family didn't have answers. Wanda lit up with wonder.

Wondering meant work. She looked things up in books and tried to understand. She didn't ask questions at school, though.

School was like the painful shots Wanda needed to manage her diabetes, a disease she'd had since she was little. School wasn't fun.

Sometimes Wanda and her best friend skipped class.
But when their parents caught them, they went back.

Eventually Wanda found challenges she enjoyed in school. She even entered and took second place in a local science fair competition just to prove she could.

By high school Wanda knew she wanted to go to college. There wasn't much money, but her parents told her if she worked hard, she could study anything she wanted.

Wanda wanted to understand the universe.

For two years she studied physics and aimed for the stars.

In the third year, life changed. Wanda found she couldn't read the writing on the classroom board. She managed by asking her friends for their notes.

When she started getting lost in the hallways, she walked with her friends.

Then Wanda couldn't see to unlock her apartment door.

In fact, she couldn't see much at all.
Diabetes was taking her eyesight.
Wanda was going blind.

Her roommate, Lucy, helped her with the door and told her about training so she could live independently. Wanda ignored Lucy. She didn't want to admit she needed help.

But Lucy insisted. Finally Wanda agreed.

She learned to walk with a white cane and count her steps so she wouldn't get lost. Now she could get around campus. Life was getting easier.

But the stars had dimmed, too. How could she study what she couldn't see?

It's going to be so hard, her mind whispered. *There's no way for you to do physics.*

No, Wanda answered, *I'm not going to change my course!* But she didn't know how to keep going.

Then her friend Emilio asked her to check out his project.
He turned on a radio. It made a crackling sound.

Wanda thought it was just static—the sound when
there's no radio station nearby.

Sh sh sh sh sh sh sh sh

Emilio told her no, the sounds were radio waves from space.

sh sh shshshshshsh

Wanda was not impressed.

"Wait," said Emilio.

And then . . . ¡Cambió! It changed!

sh sh sh SHSHSHSHSH

For Wanda it was like a giant wave crashing on a beach. She could hear the sound transform.

"That's a solar burst," Emilio said. The radio was picking up an explosion on the Sun!

"¡Fascinante!" said Wanda. In that moment, she heard a path to the sky.

Now that Wanda knew how to get there, she flew.
She got into an internship program at NASA's
Goddard Space Flight Center in Maryland,
far from Puerto Rico.

"I was terrified!" she said later. But with the help
of a guide, she learned her way around.

Counting steps, she walked from the bus stop to her office. There she used sonification to listen to what other people looked at.

Tum-tum-tumooooTUM

A computer turned data into sounds like drumbeats and chimes. Wanda listened for patterns in the sounds. She was finally studying the stars!

But she still worried. Could she do as much as the other scientists?

To find out, Wanda tested herself.

She listened to data from an exploding star and wrote what she thought was happening. Another scientist looked at graphs and tables of the same data. Then he wrote up his conclusions.

The results were the same.

"¡Victoria!" Wanda yelled. Sonification worked! And she had even heard something new: oscillations, or tiny waves, in the energy from the star. No one had noticed them before.

"There's so much I can hear in a starbeat," Wanda said, explaining her work to others. Using sound didn't mean she had less information. It just meant she had to work differently.

Wanda kept going to school and finished her doctorate. Today she shares her work with fellow scientists, people with disabilities, and anyone else who will listen. People flock to her talks.

"Science is for everyone!" she says. And then she invites her audience on a galactic audio tour.

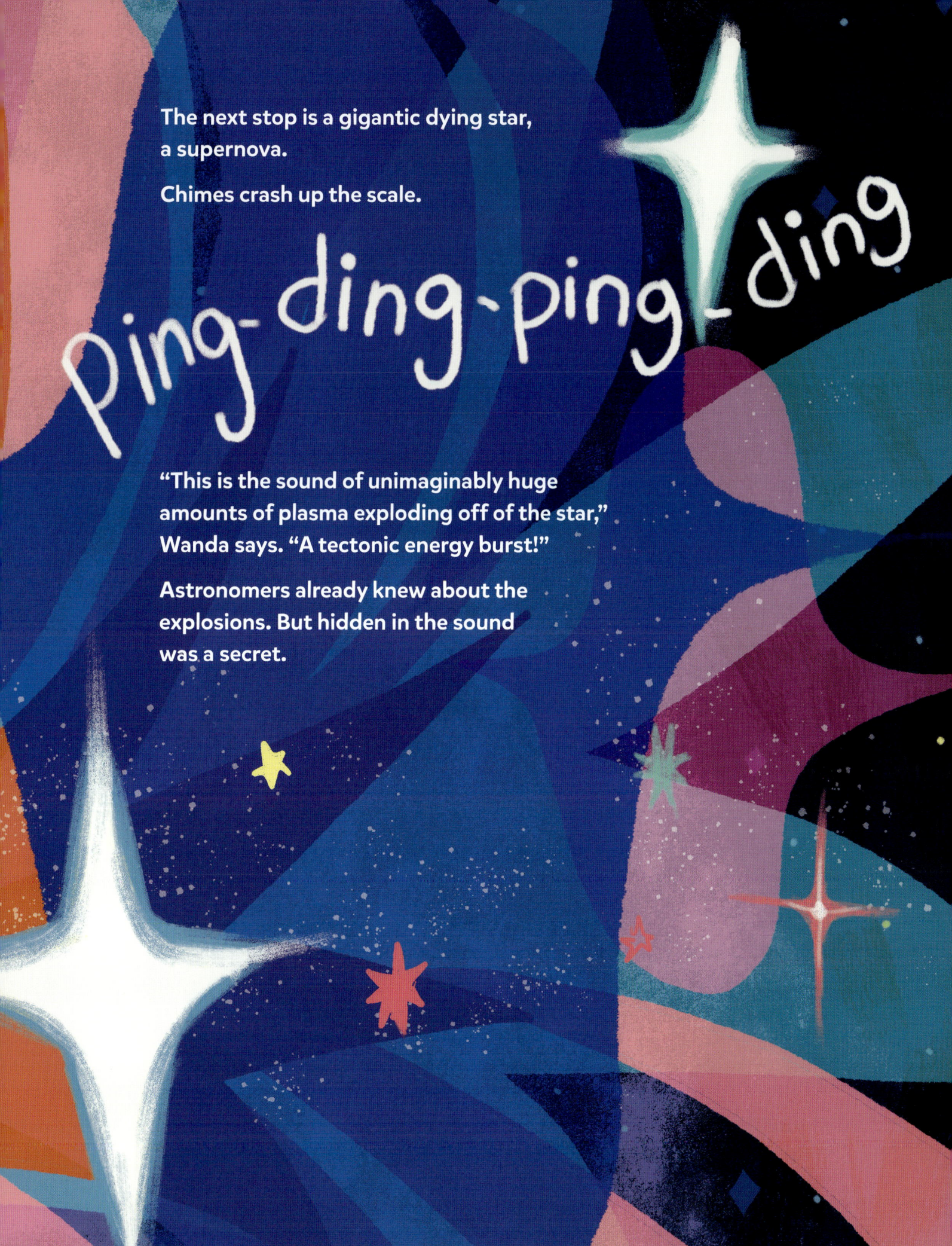

The next stop is a gigantic dying star, a supernova.

Chimes crash up the scale.

Ping-ding-ping-ding

"This is the sound of unimaginably huge amounts of plasma exploding off of the star," Wanda says. "A tectonic energy burst!"

Astronomers already knew about the explosions. But hidden in the sound was a secret.

Wanda slows the tones down.

poodah poodah poodah

The audience hears waves of energy caused by those plasma bursts.

Wanda and other scientists listened to the energy waves. They found that a dying star's energy sets off the birth of new stars— the universe's ultimate act of recycling.

This was the secret hidden in the starbeat.

Wanda found her path to the stars by using sound. Other people will need different tools. But she knows that when everyone finally has access to the skies, she will hear

Ping-ding.

ping·ding

—an explosion louder than a supernova. And this time, she says, the explosion will be a "tectonic burst of knowledge."

Look up! ¡Mira arriba!
The stars belong to everyone!

Glossary

braille: A system of reading and writing that uses six raised dots. The reader uses their fingers to read the words.

diabetes: Type 1 diabetes, which Wanda developed as a child, is an autoimmune disease that attacks the pancreas so the body doesn't make the insulin it needs. People with type 1 diabetes take insulin shots.

meteor: A rock from space that hurtles into Earth's atmosphere and burns up. A meteor is sometimes referred to as a falling star because it glows brightly as it burns, but it is not actually a star.

plasma: One of the four states of matter: solid, liquid, gas, and plasma. Stars create plasma by heating gas to super high temperatures.

radio waves: A type of invisible light energy. Radio waves are emitted by many things in space, including stars, black holes, and planets. These radio waves can be picked up by radio receivers on Earth.

sonification: The presentation of data through sound.

slate and stylus: A slate and stylus are to a braille reader what a pencil or pen is to a print reader. The blind or visually impaired person uses the pointed stylus to make raised dots on paper. The slate helps position the dots accurately.

supernova: A dying star that is in the process of exploding. Neutron stars are the gigantic stars that become supernovae when they explode.

tectonic: Huge or earth-changing.

Spanish Terms

¡Cambió! (kahm-bee-OH): It changed!

¡Los colores! (LOHS koh-LOH-rays): The colors!

¿De qué están hechas las estrellas? (deh KEH es-TAHN EH-chahs LAHS es-TREH-yahs): What are stars made of?

Esas estrellas son piedras que caen del cielo (EH-sahs es-TREH-yahs SOHN pee-EH-drahs KEH KAH-en DEL see-EH-loh): Those stars are stones that fall from the sky.

¡Fascinante! (fah-see-NAHN-teh): Fascinating!

¡Mira arriba! (MEE-rah ah-REE-bah): Look up!

¿Qué están haciendo en el cielo? (KEH es-TAHN ah-see-EN-doh EN EL see-EH-loh): What are they doing in the sky?

¡Victoria! (vik-TOH-ree-ah): Victory!

A Note from Wanda

Growing up on the beautiful island of Puerto Rico, I experienced many sunsets. To sighted people, a sunset resembles the Sun sinking into the water. For me, all of that visual grandeur turned off over time. Today I perceive the world in a very, very different way.

I was working on my science studies in my early twenties when my sight loss accelerated dramatically. Eventually I lost my sight completely.

I had so many good mentors. They told me, "Do not give up!" But sometimes your mind can be your worst obstacle. It said, *If smart people haven't done it, it's because it is impossible*. I fought back and said, *No, I'm going to do it*.

Eventually I learned that being blind doesn't mean that I have to miss the spectacle of the sunset or stop doing science. No, the spectacle and the science are better! Now I hear things scientists can't see! I do this using data collected from telescopes on the ground and in space. The telescopes measure different types of light. Physicists convert the measurements into numbers. Often those numbers are made into images and graphs. For me, the numbers are changed into sound.

That is the process of sonification, and it is how I listen to the stars. Listening to those measurements is like diving into the very heart of the universe!

My journey has led me to friends such as Amy. I met Amy at a time when it felt almost impossible to reach my goals. But there was a way! I emerged with what I call "unbeatable determination."

Working on this book with Amy, I've tried to use what has happened to me to encourage you to keep trying! Success does not happen overnight. It requires dedication. Don't give in to discouraging thoughts. If there's not a way, create one yourself, one step at the time, little by little. Never give up!

Wanda at the International Astronomical Union's General Assembly in Vienna, Austria. Credit: IAU/M. Zamani.

"Science is for everyone! It belongs to all people because we are all natural explorers."

—Wanda Díaz Merced

A Note from Amy

I met Wanda when she first came to Maryland. She was an intern assigned to my husband's office at NASA's Goddard Space Flight Center. We had a dinner party that summer, and Wanda's energy filled the room.

Wanda and I spent some time together in the summers. I tried listening to the data that she was working with. It was astounding. If I were going to study the data the way Wanda and other scientists were doing, my brain would need to be trained. And that, I realized, was how she was working. She was training her brain, just as younger students train their brains to learn to read and write.

Amy and Wanda at NASA's Goddard Space Flight Center. Credit: Robert Candey.

Our family visited Wanda in Puerto Rico. I visited her in Scotland while she was doing her PhD work and again in Boston when she was doing research at Harvard.

In early 2020 Wanda and I agreed to write a book together. I made plans to take a break from work and visit her in Colorado. Then the pandemic hit. So instead of spending an intensive week together, we set up a call several times a week. We kept going for months. We talked about our families, we talked about books we were reading, and, of course, we talked about her journey. In some cases I asked her to try to reconstruct what people said to her. We recreated many of the sentences in Spanish to better capture the scene. I also read Wanda's published papers, watched her videos, and talked with people who worked with her.

While Wanda doesn't wear a cape, she is certainly a super woman. Her energy and determination have not only overcome many setbacks but also led to success in her field.

Wanda Tidbits

When she knows where she's going, she walks very fast.

She laughs a lot.

She wants to know everything about everything, and her memory is phenomenal.

She uses a lot of exclamation points because life is exciting!!

Quotation Sources

"No, I'm not . . .": "Listen to the Stars" TEDx Talk, July 2, 2014.

"Science is . . . ," "This is the sound . . . ," and "tectonic burst . . .":
"How a Blind Astronomer Found a Way to Hear the Stars"
TED Talk, July 13, 2016.

"Astronomy is responding . . .": UN address, February 11, 2022.

All other quotations come from Amy and Wanda's conversations.

Radio JOVE

sh sh sh SHSHSHSHSH

Jupiter's atmosphere usually sounds like waves crashing on a shore.

NASA's Radio JOVE education program wants everyone to know about these sounds. Students and citizen scientists build receivers and antennas from kits. They record their observations on a website.

"All planets have some sort of radio waves," explains project scientist Leonard Garcia. Radio JOVE focuses on Jupiter because that planet's waves are stronger and more easily heard than those of other planets.

The atmosphere on Jupiter has a lot of electrical storms. These storms send waves of energy, including radio waves, into space. When the radio waves reach Earth, they can be picked up by Radio JOVE receivers.

When Wanda got to NASA's Goddard Space Flight Center as an intern, she wanted to build a Radio JOVE receiver kit herself.

"It was a remarkable thing that I never thought would work," said Leonard. How could Wanda manage the soldering without seeing any of the pieces? "But she was saying, 'Hey, I want to take on this project!' So we figured out how to do it."

Leonard read the instruction manual aloud and handed out the parts in the right order. Wanda and the other interns each built their own receiver. Eventually they took their work outside to an antenna.

Hooking up the receivers, they listened: *shshshSHSHSHshshshSHSHSH*. They heard Jupiter!

Wanda was able to explore space with the radio kit, just like others use telescopes. Now when Wanda travels, she takes radio kits with her so she can show her students that they, too, can reach for the stars.

Learning to Get Around

In college, when Wanda was losing her sight, an orientation and mobility specialist worked with her until she could confidently get around on her own. Later, when Wanda got to NASA's Goddard Space Flight Center, she met Disability Coordinator Denna Lambert. Denna's job is to make sure employees and interns have what they need to succeed.

Wanda and Robert Candey at NASA's Goddard Space Flight Center. Credit: NASA/Deborah McCullum.

Wanda needed tools. Denna gave her a braille protractor, a tactile map, and the paper she needed for her slate and stylus when she wanted to do math by hand. When Wanda was ready to walk around campus on her own, Denna helped her find her way. Starting from the bus stop, they practiced going to Wanda's office, to the cafeteria, and to the health unit. Denna, who is visually impaired herself, encouraged Wanda to walk confidently.

Denna also introduced Wanda to other employees who are blind or visually impaired. "Wanda can make a friend out of anyone," says Denna. "I tried to make her know she wasn't alone." After the first year, Wanda joined Denna when students from the National Federation of the Blind came to visit for the day. Today, Denna and Wanda continue to find ways to introduce students to the stars.

Wanda's Journey as a Scientist

1997: Wanda enrolls at the University of Puerto Rico at Río Piedras. Due to diabetes, she loses her sight completely while at the university. She decides to keep studying, repeating classes until she gets her degree.

2003: After six years, Wanda earns her bachelor of science degree in physics.

2005–2010: During the summers, Wanda interns at NASA's Goddard Space Flight Center and works on sonification with her mentor, Robert Candey. She studies events in space even as she works with experts to develop tools that fit her needs.

2006: Teatro Sol y Luna (Theatre of Sun and Moon) in San Juan, Puerto Rico, produces a play that uses Wanda's sonification work for a dance scene.

2007: Wanda receives a master's degree in education from the University of Massachusetts Boston.

2013: Wanda finishes a doctorate in computer science at the University of Glasgow, UK, where she studies space data analysis.

2013: Wanda accepts post-doctoral fellowships at the Harvard-Smithsonian Center for Astrophysics in Cambridge, Massachusetts, and the South African Astronomical Observatory in Cape Town. She presents talks all over the world, sometimes asking her audience to close their eyes so they can experience space as she does.

2013: Gerhard Sonnert, a colleague at the Center for Astrophysics, publishes music called "Star Songs," inspired by Wanda's work.

2014: Wanda presents her TEDx Talk, "Listen to the Stars," at Westerford High School in South Africa.

2016: Wanda presents her TED Talk, "How a Blind Astronomer Found a Way to Hear the Stars," in Vancouver, Canada.

2016: Wanda is invited to the White House Frontiers Conference hosted by President Barack Obama in Pittsburgh, Pennsylvania.

2017: The BBC names Wanda to its "100 Women" series on trailblazing women in science. She is listed alongside Nobel laureate Marie Curie.

2018–2019: Wanda works for South Africa's Office of Astronomy for Development. She advocates for the use of multisensorial data, which gives all scientists access to information.

2018–2019: Wanda and her colleagues at the International Astronomical Union present *Inspiring Stars*, a traveling exhibition on inclusive astronomy projects.

2020: The United Nations (UN) adopts a resolution supporting the inclusive approach to sciences.

2022: Speaking at the UN, Wanda says, "Astronomy is responding to the urgent need to integrate all performance styles." She is confident this integration has the power to "scientifically address human needs and rewrite human history."

2023: Wanda joins the faculty at Universidad del Sagrado Corazón in Puerto Rico.

Explore More

100 Women: Seven Trailblazing Women in Science: https://www.bbc.com/news/science-environment-41861232
Read about Wanda and other women whose work is pushing the frontiers of science.

"How a Blind Astronomer Found a Way to Hear the Stars | Wanda Díaz Merced": https://www.youtube.com/watch?v=-hY9QSdaReY
Watch Wanda's TED Talk about her journey.

"Listen to the Stars: Wanda L Díaz Merced at TEDxWesterfordHighSchool": http://www.youtube.com/watch?v=wbtLTCA1Qd4
Watch Wanda's TEDx Talk for high school students.

The Mysteries of the Universe **by Will Gater (DK Children, 2020)**
Read more about astrophysics, Wanda's field of study.

The Radio JOVE Project: https://radiojove.gsfc.nasa.gov/index.php
Join NASA's Radio JOVE Project. Listen to the Sun and Jupiter and build your own radio telescope kit.

Wanda Díaz Merced: The Astronomer Who Hears Stars: https://www.beyondcurie.com/wanda-diaz-merced
Explore a design project celebrating Wanda and other women in STEM.

What Are Radio Waves?: https://www.nasa.gov/directorates/heo/scan/communications/outreach/funfacts/what_are_radio_waves
Learn more about radio waves and the electromagnetic spectrum.

Selected Bibliography

Adriaanse, Dominic. "Innovation Helps Blind Enjoy Exhibits." *Independent Online.* Jan. 21, 2016. https://www.iol.co.za/capetimes/news/innovation-helps-blind-enjoy-exhibits-1974307.

D'Antonio, Maria Rosaria, Lina Canas, and Wanda Díaz Merced. "Inspiring Stars: The IAU Inclusive World Exhibition." *Proceedings of the International Astronomical Union* 13, no. S349 (Dec. 2018): 470–473. doi:10.1017/s1743921319000620.

Díaz Merced, Wanda. "Sound for the Exploration of Space Physics Data." PhD thesis, University of Glasgow, 2014. https://eleanor.lib.gla.ac.uk/record=b3090263.

———. "The Sounds of Science." *Physics World* 24, no. 6 (June 2011): 42–43. doi:10.1088/2058-7058/24/06/41.

———. "Wanda Díaz Merced at the United Nations—IDWGS2022." EGO & the Virgo Collaboration. Streamed live on Feb. 14, 2022. YouTube video, 5:27. https://www.youtube.com/watch?v=Mho8uryWcsM.

Díaz Merced, Wanda, and Michael Gastrow. "Astronomy and Inclusive Development: Access to Astronomy for People with Disabilities." *Proceedings of the International Astronomical Union* 14, no. A30 (Aug. 2018): 596–597. doi:10.1017/s1743921319005593.

Díaz Merced, Wanda, et al. "Exploring Sound to Convey Information." Poster presented at the 218th Meeting of the American Astronomical Society, Boston, MA, May 2011.

Garcia, Beatriz, Wanda Díaz Merced, Johanna Casado, and Angel Cancio. "Evolving from xSonify: A New Digital Platform for Sonorization." *EPJ Web of Conferences* 200 (Feb. 1, 2019): 01013. doi:10.1051/epjconf/201920001013.

Gibney, Elizabeth. "Q&A Wanda Díaz Merced." *Nature* 577 (Jan. 9, 2020): 155.

Gonzalez-Espada, Wilson Javier. "Listening to the Whispers from the Stars." Ciencia Puerto Rico. Oct. 1, 2013. https://www.cienciapr.org/en/monthly-story/listening-whispers-stars.

Hendrix, Susan. "Summer Intern from Puerto Rico Has Sunny Perspective." NASA. Last modified Apr. 28, 2011. https://www.nasa.gov/centers/goddard/about/people/Wanda_Diaz-Merced.html.

Johnson, Lisa. "Blind Astrophysicist Listens to the Stars by Turning Data into Sound." CBC. Feb. 18, 2016. https://www.cbc.ca/news/canada/british-columbia/star-sounds-wanda-diaz-merced-ted-1.3452236.

Jones, Graham, and Richard Gelderman. "Listening to the Patterns of the Universe." EarthSky. Nov. 29, 2018. https://earthsky.org/space/space-data-into-sound-patterns-wanda-diaz-merced.

Kurtz, S., et al. "High Resolution Radio Continuum Observations of High Mass Star Formation Regions." *Symposium: International Astronomical Union* 205 (2001): 280–281. doi:10.1017/s0074180900221219.

Sonnert, Gerhard. "X-Ray to Sound: A Fortuitous Accident." Star Songs: From X-Rays to Music. 2012. https://lweb.cfa.harvard.edu/sed/projects/star_songs/pages/xraytosound.html.

To my family and friends, who have helped so much.—A. S. H.

To my mom, to my sister, to my Puerto Rico, to my mentors: Daisaku Ikeda, Robert Candey, Dr. Nancy Brickhouse, Professor Matthew, and Dr. Katsanevas!—W. D. M.

To my son, the star of my life. May you always find your light in adversity.—R. A. M.

Charlesbridge • 9 Galen Street, Watertown, MA 02472 • www.charlesbridge.com

Library of Congress Cataloging-in-Publication Data
Names: Hansen, Amy, author. | Díaz Merced, Wanda, author.
  | Arreola Mendoza, Rocío, illustrator.
Title: Wanda hears the stars / Amy S. Hansen with Wanda Díaz Merced; illustrated by
  Rocío Arreola Mendoza.
Description: Watertown, MA: Charlesbridge, [2025] | Includes bibliographical references.
  | Audience: Ages 6–9 | Audience: Grades 2–3 | Summary: "Growing up in Puerto Rico,
  Wanda Díaz Merced wanted to study the stars. But when she lost her sight, she had to
  find a new way to work. Through the use of sonification, which turns data into sound,
  she was able to make a path for herself and other scientists with disabilities."—
  Provided by publisher.
Identifiers: LCCN 2023056535 (print) | LCCN 2023056536 (ebook) | ISBN 9781623544874
  (hardcover) | ISBN 9781632894298 (ebook)
Subjects: LCSH: Díaz Merced, Wanda–Juvenile literature. | Blind astronomers–Puerto
  Rico–Biography–Juvenile literature. | Women astronomers–Puerto Rico–Biography–
  Juvenile literature. | Scientists with disabilities–Puerto Rico–Biography–Juvenile
  literature. | Astronomy–Data processing–Juvenile literature. | Computer sound
  processing–Juvenile literature. | Sound in astronomy. | LCGFT: Biographies.
Classification: LCC QB36.D49 H36 2025 (print) | LCC QB36.D49 (ebook) | DDC 520.92
  [B]–dc23/eng/20240109
LC record available at https://lccn.loc.gov/2023056535
LC ebook record available at https://lccn.loc.gov/2023056536

Printed in China • OPIC
The authorized representative in the EU for product safety and compliance is eucomply
  OÜPärnu mnt 139b-14, 11317 Tallinn, Estonia, hello@eucompliancepartner.com, +33757690241
(hc) 10 9 8 7 6 5 4 3 2

Illustrations done in digital media
Text type set in Mikado by Hannes Von Döhren
Edited by Alyssa Mito Pusey with assistance from Natalia Vázquez Torres
Designed by Diane M. Earley
Production supervised by Nicole Turner